Cyber Security

Incident Response Plan

By

Travis Lothar Czech

Dedication

To everyone who has given me a chance in work and education—your belief in me has made all the difference. To my friends and family, whose unwavering support and encouragement have been the pillars upon which I have built my dreams. And, most importantly, to Lance Thorpe, a student who has taught me so much through our mentorship and coaching journey. Thank you, Lance, for showing me the true meaning of learning and growth, and for inspiring me to write this book. This one's for you.

Table of Contents

Introduction
Don't Panic! The Hitchhiker's Guide to Cyber Incidents

Yo, (Open up, man) What do you want, man?
(My manager just caught me) You let her catch you?
(I don't know how I let this happen) With what?
(The computer over there, you know) Man.
(I don't know what to do) Say it wasn't you.
(Alright)
Manager came in and she caught me red-handed.
With the USB stick I found on the floor.
Picture this, I put it in without scrubbing.

Yes, this happens far too often. You know that feeling when you stumble upon a USB stick on the ground and can't resist the urge to plug it into your computer? It's like an irresistible cyber treasure chest, just begging to be opened (you know, maybe an iCloud leak of Caiti Jerry). You walk into the office, shoulders back and head held high. Just imagining what could be on the USB stick. You sit down at your workstation PC, look left-then right. A grin forms on your face, and you stick it in without getting it checked or any sort of protection.

But then, in a twist of fate worthy of a Hollywood thriller, your computer starts acting up. Internet security alerts pop up, the screen turns black, and suddenly you're the starring character in your very own cyber horror movie.

Oh no! your face goes devoid of colour as all your blood rushes from your face and your heart races as you wonder what sort of digital monster you've just unleashed and your flight or fight response kicks in...

Then the manager walks by and sees that the FBI has seized your workstation. Your escape route is blocked as the entire office is drawn by the audible *"This operating system is locked due to the violation of federal law..."*

This is what we call an incident, and this is what brings you, the reader to this book.

It isn't enough to have policy in place to stop incidents from happening. Let's be honest, how many of us actually read information technology (IT) policies or terms of service (ToS) agreements. You can run education campaigns day and night. Not everyone will listen to them or even retain the *Death by PowerPoint* that we are all too familiar with.

So, in response to that—we have Incident Response Plans (IRPs). We'll get into more detail of exactly what that is. But, think about it like, "O noooooooooo! Something happened, what ever should I do?".

Close your eyes and picture a pale white freckled nerdy guy. God rays, and raytracing are enabled, and you are blinded by the super-hero aura of the IT guy with the IT bible in his hands, "I know what to do!" he says as his frail body trembles.

He runs through the immediate actions (IAs) drills with the proficiency of a tenured military veteran and his C7 rifle. You stand there in awe of his majesty as he operates this piece of digital machinery with such ease as you could have mistaken it as an additional appendage (yes, a keyboard is akin to an arm for an IT professional).

What was in that IT bible you might ask? That is every policy, and procedure that this God amongst digital men has probably written and you've never read. But it has that IRP in it for ransomware.

What are they doing with that policy? Saving you and your organization. Okay, maybe not that much. Depending on the nature of the infection it could contain advice from your lawyer (data breach) or other IT professionals such as a cyber security analyst.

So, let's get to the content of this book. We'll start off by covering some of the basics such as the OSI model, CIA triad, and the Cyber Kill Chain. Once we're done with some of the basic theory, we are going to cover some of the Cybersecurity Frameworks to reference when building out policy. We'll then move on over to types of malware and how they affect the system. Once we are done with that we will move into your initial reaction and responses. We won't actually create policies and procedures. However, we'll cover different types of them. Finally, we'll get into IRPs and how to write them.

So, buckle up, grab your favorite beverage, and get ready for a wild ride through the fascinating, sometimes terrifying, but always entertaining world of cyber incident response. Let's dive into Chapter One: The Wonderful World of OSI, CIA Triad, and Cyber Kill Chain. Trust us, it's way more exciting than it sounds, and you'll come out the other side feeling like a true cyber warrior.

Chapter 1: Cyber Theory
The Good, the Bad, and the Just Plain Weird

Introduction to Cyber Security

Okay, so in the first book, we covered most of this. I'm going to focus on the things that really matter to this subject. The reason why I want to hammer down the theory of these three topics is because it will make it easier to break down the subjects. The OSI model is a standard that every field dealing with systems (computer systems) knows and understands. This will allow you to deal with engineers, scientists, and even business technology firms. Then the CIA triad helps you explain the balance between the three topics (confidentiality, integrity, and availability). Translating this to customers, clients, and even government agencies will allow you to better speak their languages. The cyber kill chain is super important for this book, as it will allow you to figure out what stage of the attack you have detected and how to respond to it.

OSI Model
A Layer Cake of Cybersecurity

You might recall that the OSI Model (Open Systems Interconnection Model) is a mouth-watering seven-layer cake that describes the functions of a networking system. But instead of delicious frosting, we're talking about all the things that can go wrong. Yikes! So, let's take a stroll through this cake and see what devious threats are lurking at each layer.

5

7. Application Layer
- SMTP // FTP // HTTP

6. Presentation Layer
- NCP // TDI // AFP

5. Session Layer
- SSL // TLS // NetBIOS

4. Transport Layer
- TCP // UDP

3. Network Layer
- Routers // IP Protocol

2. Datalink Layer
- Network Switches

1. Physical Layer
- Hubs // Repeaters

We're going to focus on the threats that are targeting each layer. We can then build on that knowledge for future testing and creating our Incident Response Plans (IRPs). So just imagine you're looking down the barrel of a gun. You're sighting on your target. With my background, I'm used to seeing the *NATO Figure 11* target. So that's what I picture when I think of a target.

Application Layer: Ah, the top of the cake. Layer 7 is where all the fancy stuff happens. This is where we have vulnerabilities such as Distributed Denial-of-Service attacks (DDoS), Hyper-Text Transfer Protocol (HTTP) floods, Structured Query Language (SQL) injections, Cross-Site Scripting (XSS) attacks, and the list goes on. It's like a smorgasbord of cyber threats, and it's hard to protect because there are just so many ways users can interact with applications. They target vulnerabilities in the application layer itself (It's like trying to keep a toddler away from a freshly baked cake - nearly impossible!). It is honestly the hardest to defend against. The reason is that there are numerous ways for you-the user, to input information. The application is in the hands of the user, which makes it highly complex. Tools such as anti-cheat software have been defeated countless times for similar reasons. We'll expand on this topic in *Chapter 2* but for now, we will move onto the sixth layer.

Presentation Layer: Down to Layer 6, and it's time to talk about the sneaky exploits that can happen here. Man-in-the-Middle (MitM) attacks, SSL hijacking, and SSL sniffing are just a few of the mischief-makers you might find. It's like having a stealthy ninja lurking in the shadows, waiting to pounce. But fear not, aspiring analyst; we'll shine a light on these evildoers later on.

Session Layer: we've got the presentation layer out of the way and are on the Layer 5 is where we deal with encryption and user sessions. If cyber threats were a rock band, MitM, Sniffing, Session Hijacking, and Session Downgrade Attacks would be the supergroup. Now that we are done here, we can head over to the transport layer.

Transport Layer: Zooming in on Layer 4, we find some lovely encryption happening with Transport Layer Security (TLS). But, alas, there are villains here too, like SYN flooding, TCP Blind Spoofing, and Connection

Hijacking. It's like a game of Whac-A-Mole, but we're going to whack these moles with knowledge! You'll even notice that most of these attacks are related to the TCP protocol.

Network Layer: Welcome to Layer 3, home of the notorious Internet Protocol (IP) Spoofing, Smurf attacks, and Ping of Death. These cyber baddies are like the mean kids on the playground, but we're going to stand up to them by learning all about their dirty tricks. Things such as Address Resolution Protocol (ARP) Poisoning, Traffic Sniffing, Byzantine Attack, Port Stealing, or Routing Attacks. There are many more types of attacks we can talk about before moving onto *Layer 2*. But you get the gist—this is a layer you can exploit.

Data-link Layer: Layer *2*, a word of advice *Don't click any suspicious links*, they may steal your data. With the data-link layer being so secure, what could possibly be done to exploit it? Well, an attacker can: leverage Media Access Control (MAC) Spoofing/Cloning or Flooding, broadcast storms, and other Frame-level exploits. But I think it's time to get physical with the *Physical Layer*.

Physical Layer: Layer *1*, is the final layer. You might be thinking *"But, hackers are keyboard warriors"*. However, this isn't always the case—they can use physical exploits. Just think of *Action Replay Cards* or pulling a Complementary Metal-Oxide Semiconductor (CMOS) battery to bypass an admin password. With threats like wiretapping, port access control violations, and even good old-fashioned theft, this layer is like the seedy underbelly of the cyber world. Think of it as the dark alleyway where bad guys lurk, waiting to pounce on unsuspecting victims. But don't worry; we'll learn how to stay street-smart and keep our precious hardware safe. Basically, anything that can be thought of being physical falls in this category for attacks. But alas we must move on.

Some people argue SSL is on Layer 5, and others argue Layer 6. So just be mindful when talking to others or taking any level of certification in the field—that you have to know what the instructor/examiner is expecting.

Now that we've had a whirlwind tour of the OSI model's seven layers, let's move on to our next cyber concept: the CIA triad. Don't worry; it's not about undercover spies (although that would be cool, right?). This triad is all about keeping our data safe and sound.

CIA Triad
No, It's Not About Espionage!

Now, we have two words here *CIA* and *Triad*. No, I don't mean the *Central Intelligence Agency* or the *Crime Syndicate*. What I am referencing is the relationship between *Confidentiality, Integrity, and Availability*. It's like the holy trinity of cybersecurity. These three principles form the foundation of a robust security strategy. Think of them as the three legs of a stool; if one leg is weak, the whole thing wobbles.

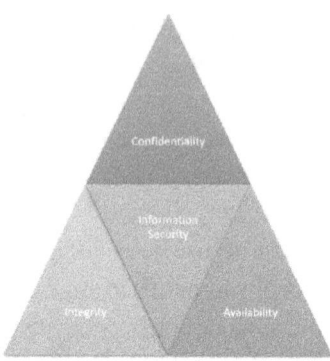

Basically, what this is representing is risk analysis or a balance. You have weight each variable and classify data accordingly. Think about it if everyone in your country needed to have a *Top-Secret* security clearance just to know each other's name. Otherwise, everyone was nameless. Well, that would be an over classification, the information's integrity would be maintained, but it isn't very available. Now how could you pick up at a bar. Hmmm, I guess that could be a good pick-up line. *"Hey, what's your name □ "*, *"If I told you, I'd have to kill you -_-"*, *"I think it's worth the risk □ "*. However, I think those lines only work in the movies. Not all of us are *James Bond*.

Confidentiality: Most of us probably think of when we are protecting systems or digital property—it's to protect confidential information. Encryption, access controls, and authentication are just a few of the ways we can ensure that only authorized eyes can see our sensitive data. It's like having a secret handshake or a members-only club. Think about it, do you want everyone to know a company or government's security plans? Would you want everyone to know your PIN for your bank? Probably not!

However, there is the possibility to over-classify information which would reduce the ability to access it. Under, and over classification could cause problems. Let's think about it from another frame of reference. Imagine you are a politician, and you didn't want the public to know something—like you eliminated or extorted an opponent, so you deemed those communications *Secret* and suddenly anyone who leaked it would be charged. The public wouldn't know you're dirty, but it's still malicious.

On the other hand, you could under classify something like ongoing battle maneuvers in a warzone. Suddenly the public and your enemy are watching your attack plans on *CNN*—whoops, you just put all the soldiers' lives in jeopardy because of a mistake.

Integrity: Integrity is all about making sure that our data remains unaltered and uncorrupted. It's like having a trusted friend who will always tell you the truth, even if it's hard to hear. Let's look at this from a financial standpoint. You want to keep financial records safe, secure, and accurate—however, you the customer still need to be able to access them. Integrity is essential here because you want to know for certain how much money you have, and the bank wants to make sure your transactions are correct. The integrity of the bank ledger is important. This is also why blockchain is such a great technology—through consensus you ensure the integrity of the blockchain. Then we have something called availability.

Availability: Finally, availability is all about ensuring that our data is always accessible when needed. This means keeping our systems up and running, safe from crashes and downtime. This is where things get interesting. I think so anyways. Because availability is the key to information. It's one thing to have information, to store it and hoard it. But you need to be able to use that information. You can't use data if it is locked up—the key is thrown away. However, you don't want the wrong people to find it—abuse it.

Basically, when it comes to data and data protection, you are performing a balancing act between these three things. You need to limit accessibility to who should have access to it and the level of access to the information.

Think about *NT (New Technology) File System (NTFS)* and the permissions.

Permissions for SYSTEM	Allow	Deny
Full control	✓	
Modify	✓	
Read & execute	✓	
List folder contents	✓	
Read	✓	
Write	✓	

You've got a bunch of permissions. You can give someone full control, modify, read & execute, read, write, and a couple more. So, your admins should have a high level of permissions—but, should someone who only needs to read files require the same level of access as a *System Administrator*? Probably not.

Another thing to think about is a process to strip rights away from someone who gets fired or quits. Just imagine, *Tom Smykowski* from *Office Space* getting fired, but he still has access to the office network. He decides he is going to download company secrets and get a nice payday so he can retire young. Well, IT should've had a policy to prevent that!

Situations like that and things like *Canadian Revenue Agency* or *Ashley Madison* data leaks are examples of why data protection is essential.

Now it's time to move on over to the *Cyber Kill Chain!*

Cyber Kill Chain
Kill Chain, or Buzz Kill?

The cyber kill chain is a seven-step process that describes how an attacker infiltrates, exploits, and ultimately compromises a target system. It's like a play-by-play breakdown of a cyber heist. To help you better understand each stage, we'll walk you through a hypothetical cyber-attack using the kill chain framework and provide examples for each step. So, buckle up and get ready for a wild ride through the dark side of cyberspace.

Here we have:

1. Reconnaissance: In this stage, the attacker gathers intel on their target, like a cyber detective (or stalker, depending on how you look at it). They may use social engineering tactics, such as researching employees on LinkedIn or browsing a company's website for information about its technology stack. For example, an attacker might

learn that a company uses a specific software program, which has a known vulnerability.

2. Weaponization: Once the attacker has collected enough intel, they move on to weaponization. This stage involves creating a malicious payload designed to exploit the target's vulnerabilities. For instance, the attacker might craft a phishing email with a malicious attachment or link that targets the known vulnerability in the company's software.

3. Delivery: In the delivery stage, the attacker sends the payload to the target, like a devious mailman. They might use a variety of techniques to deliver their malicious payload, such as sending phishing emails, setting up fake websites, or infiltrating the target's network using compromised credentials. In our example, the attacker sends the phishing email to multiple employees at the target company, hoping someone will fall for the bait.

4. Exploitation: The payload exploits a vulnerability in the target system, like a master lock-picker. If the target takes the bait (e.g., opens the malicious attachment or clicks the link), the payload executes and exploits the known vulnerability. In our example, one employee falls for the phishing email and clicks on the malicious link, allowing the attacker's payload to exploit the software vulnerability.

5. Installation: Once the exploit is successful, the attacker installs malware on the target system,

like a sneaky burglar setting up shop. This malware can take many forms, such as ransomware, spyware, or Trojans, and often has the ability to propagate throughout the target's network. In our scenario, the attacker installs a remote access Trojan (RAT) on the employee's computer.

6. Command and Control: After installing the malware, the attacker establishes a remote connection to the target, like a puppet master pulling the strings. The attacker now has control over the compromised system and can use it to carry out further attacks or gather additional data. In our example, the attacker uses the RAT to gain access to sensitive company information and harvest login credentials for other systems.

7. Actions on Objectives: Finally, the attacker achieves their goal, whether it's stealing data, disrupting services, or causing havoc. The ultimate objective depends on the attacker's motivations and desired outcomes. In our hypothetical attack, the attacker exfiltrates sensitive company data, sells it on the dark web, and uses the stolen login credentials to launch further attacks on the company's partners.

In a nutshell the seven stages are a highly simplified variant on battle procedure or any other planning process. This is slightly expanded when talking about the eight stages. Remember, we're just using the bank robbery analogy for educational purposes, so don't get any funny ideas!

- Reconnaissance
- Intrusion
- Exploitation
- Privilege Escalation
- Lateral Movement
- Obfuscation
- Denial of Service
- Exfiltration

Reconnaissance: I like to think of this as intelligence gathering or starting your plan. Think about it—if you are going to rob a bank (what child hasn't thought of the perfect bank robbery plan) you need to research it. So, I am planning a bank robbery: floor plan would be nice, knowing guard shift change, when the most money is held at the bank, and even where the cameras are. You also, want to know your escape route and the time it takes for the police to arrive. Basically, if you want your bank heist to go well—you need intel and a plan. Same with pentesting!

Intrusion: This is the moment when our hypothetical bank robbers burst through the doors, guns blazing (or, you know, politely opening the door if they're feeling sneaky). In a cyber attack, this is when the attacker establishes a foothold in the target's network or system. For example, they might use a phishing email to trick an employee into clicking a malicious link or exploit a software vulnerability to gain unauthorized access.

Exploitation: In our bank heist, this is when the robbers start cracking the vault or forcing the teller to hand over the cash. In a cyber attack, exploitation refers to the attacker taking advantage of the vulnerabilities they've identified in the target's systems. They might use a zero-day exploit to bypass security measures or leverage a known vulnerability that hasn't been patched yet.

Privilege Escalation: This stage is all about gaining more power like our bank robbers taking the bank manager hostage to get the keys to the vault. In a cyber attack, this is when the attacker seeks to increase their level of access within the target's network or system. They might use various techniques, such as password cracking or exploiting unpatched vulnerabilities, to gain administrator or root access to the target's systems.

Lateral Movement: Here, our bank robbers are moving from room to room, looking for more loot (or hostages, depending on how ambitious they are). In a cyber attack, lateral movement involves the attacker moving through the target's network, trying to access more systems or data. They might use techniques like pass-the-hash or remote desktop protocol (RDP) hijacking to compromise additional systems within the target's environment.

Obfuscation: In the bank heist, this would be the part where our robbers disguise themselves as bank employees or customers to blend in and avoid detection. In a cyber attack, obfuscation refers to the attacker's efforts to hide their activities and maintain persistence within the target's network. They might use encryption, steganography, or other techniques to hide their malware or command and control (C2) communications.

Denial of Service: This stage is like our bank robbers setting off a smoke bomb or cutting the power to create chaos and confusion. In a cyber attack, a denial of service (DoS) or distributed denial of service (DDoS) attack involves overwhelming the target's systems or network with traffic or requests, rendering them unavailable to legitimate users. An example might be an attacker flooding a company's web server with traffic, causing it to crash and become inaccessible to customers.

Exfiltration: Finally, our bank robbers make off with the loot, high-fiving each other as they speed away in their getaway car. In a cyber attack, exfiltration is the act of stealing sensitive data from the target's network or systems. The attacker might siphon off customer data, intellectual property, or sensitive company information and then sell it on the dark web or use it for further attacks.

The Cyber Realm
A Digital Odyssey, Led by a Fearless Cybernaut

In this chapter, we embarked on our thrilling journey into the world of cyber security. We introduced the essential concepts, terminologies, and principles that form the foundation of this vast digital landscape. From defining cyber security and understanding its importance in our increasingly connected world, we explored various types of cyber threats, their consequences, and the role both offensive and defensive teams play in this high-stakes game of digital cat and mouse.

This chapter set the stage for our cyber adventure, providing a solid understanding of the field while keeping things engaging and entertaining. With our footing secured, we prepared to dive deeper into the world of cyber security in the subsequent chapters.

Chapter Two: Cyber Security Framework
A Fun-Filled Romp Through the NIST Cyber Security Framework

Introduction to Frameworks

The world of cybersecurity frameworks is just like a wacky family reunion! You've got the NIST Cyber Security Framework, but there's a whole bunch of other cousins, distant relatives, and in-laws chiming in with their own peculiar ideas. It's like being on Reddit, where everyone's got an opinion and a "brilliant" new way of doing things. And just like that quirky family gathering, sometimes Uncle ISO's method or Auntie CIS's approach might actually make sense, depending on the wild and wacky situation you find yourself in. So, buckle up, because navigating the landscape of cybersecurity frameworks is a rollercoaster ride through the land of peculiar personalities and unique insights!

Some examples of these would be:

- ISO/IEC 27001:2013 (ISO 27001)

  - This international standard provides a systematic approach to managing sensitive company information, ensuring it remains secure through an Information Security Management System (ISMS).

- CIS Critical Security Controls (CIS CSC)

- o Developed by the Center for Internet Security, this framework is a prioritized set of actions to improve cybersecurity, focusing on a series of 20 essential security controls.

- Payment Card Industry Data Security Standard (PCI DSS)

  - o Specifically designed for organizations that handle payment card transactions, this framework aims to secure credit card data and reduce payment card fraud.

- Health Insurance Portability and Accountability Act (HIPAA) Security Rule

  - o Tailored to the healthcare industry, this U.S. federal regulation outlines specific requirements for securing electronic Protected Health Information (ePHI).

- Federal Risk and Authorization Management Program (FedRAMP)

  - o Aimed at U.S. federal agencies and cloud service providers, this program provides a standardized approach to security assessment, authorization, and continuous monitoring for cloud-based services.

- European Union Agency for Cybersecurity (ENISA) Framework

  - o This European agency develops and maintains guidelines and recommendations to improve the security and resilience of national critical infrastructures and Information and Communications Technology (ICT) systems.

These frameworks, while having distinct purposes and target industries, share a common goal of providing guidance and best practices for managing and improving cybersecurity in organizations.

NIST Cyber Security Framework
Is it really just another checklist?

Buckle up and get ready to dive into the wonders of the NIST Cyber Security Framework, where we'll learn how to protect our digital goodies from the sneaky hands of cybercriminals while using the framework!

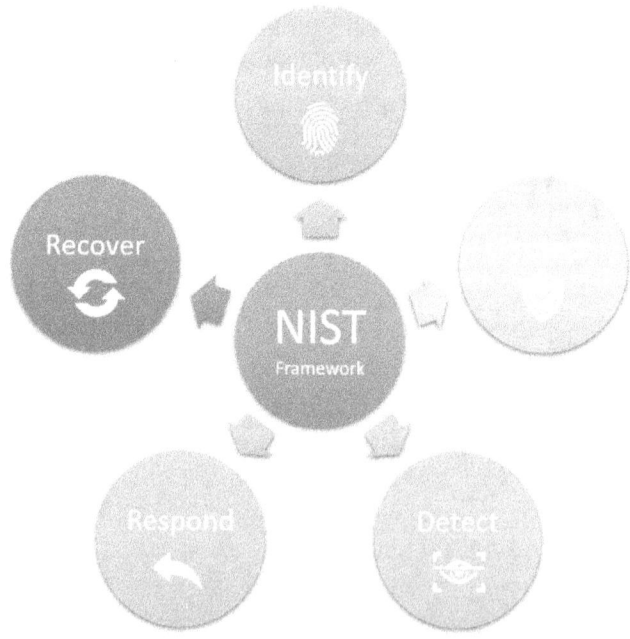

Identify: Think of this step like planning a big party (with extra emphasis on the "fun" part) at a bank. You need to make a list of everything you'll need for the party - the guests (your equipment), the food and drinks (software), and the entertainment (data). Creating a company cybersecurity policy is like setting up the rules for your bank party, such as who's invited, who's in charge of what, and what to do if someone tries to rob the bank during the festivities (i.e., dealing with a cyber attack and implementing Incident Response Planning).

Protect: This step is all about being the ultimate bank party bouncer. You control who comes in and out of your party (network access), and you make sure everyone's wearing their party hats (security software). You also take measures like encrypting sensitive data (like hiding the vault combination) and backing up your data (just in case the party gets too wild). Regularly updating security software is like checking the freshness of your banknotes, and training employees on cybersecurity is like teaching them how to detect counterfeit bills while dancing the night away.

Detect: In this step, you're the bank party detective, always on the lookout for party crashers (unauthorized access), sneaky guests bringing in their own "surprises" (unauthorized devices), or someone trying to break into the vault (unauthorized software). Investigate any suspicious activities, such as an unexpected guest who resembles a known criminal mastermind, and keep an eye on your network for any unexpected visitors.

Respond: It's time to channel your inner superhero, ready to spring into action when things get out of hand. Make sure you have a plan for dealing with potential party disasters, such as activating the alarm if there's a robbery attempt, keeping the bank operations going despite any setbacks, and calling in backup (law enforcement) if needed. Don't forget to plan for accidental mishaps, like a sudden power outage that affects your security system. Regularly testing your plan is like running a bank heist drill, but with more streamers and balloons.

Recover: This is the bank party cleanup phase. After an attack (or an epic conga line that accidentally knocked over a teller's station), you'll need to repair and restore any damaged equipment (broken cameras, compromised systems) and reassure your guests (employees and customers) that the bank is still safe and secure. It's also a time for reflection and learning, so you can make your next bank party even more secure and fun while adapting your business contingency planning. As the defender, you want to learn from the attacker's perspective so that you're better prepared for the next time they try to crash your party.

Frameworks – *Red vs Blue*
Master Chief or the Covenant

Imagine you're part of a wild west showdown between the good guys (the defenders) and the bad guys (the offensive teams). You've got three frameworks to choose from, each offering a different perspective on how to strategize and react during each phase of the battle. Let's mosey on down and compare the Cyber Kill Chain, the ATT&CK Framework, and the NIST Cyber Security Framework.

Cyber Kill Chain: This framework is like an old-fashioned, seven-stage duel between the sheriff (the defender) and the outlaw (the attacker). Each stage outlines the steps the attacker takes to plan and execute their heist, and how the defender can thwart them. For example, during the "Reconnaissance" stage, the outlaw scouts the town, while the sheriff gathers intel on the outlaw's whereabouts. By understanding each stage, both sides can anticipate each other's moves and stay one step ahead.

ATT&CK Framework: This bad boy is the comprehensive guide to every saloon brawl, bank robbery, and high-noon showdown in the wild west. It details the tactics and techniques used by the outlaws (attackers) and provides the defenders with a treasure trove of information on how to prepare and respond. Unlike the linear stages of the Cyber Kill Chain, the ATT&CK Framework takes a more holistic approach, allowing both the sheriff and the outlaw to adapt their strategies and tactics based on the evolving situation.

NIST Cyber Security Framework: This framework is like the town's rulebook for keeping everyone safe and secure. It focuses on the defenders (the townsfolk) and outlines five key steps: Identify, Protect, Detect, Respond, and Recover. By following these steps, the townsfolk can prepare for any outlaw shenanigans and minimize the impact of an attack. The NIST framework helps the sheriff rally the townsfolk and create a well-rounded defence, ensuring the town stays one step ahead of the outlaws.

Now, let's look at how both defenders and offensive teams behave and react during each phase in these frameworks.

Cyber Kill Chain: The defender's primary goal is to disrupt the attacker at each stage of the chain. For example, during the "Exploitation" stage, the attacker tries to exploit a vulnerability, while the defender works to patch it. By understanding each stage, both sides can anticipate each other's moves and counteract them effectively.

ATT&CK Framework: In this framework, both defenders and attackers can adapt their strategies based on the evolving situation. Defenders analyze the tactics and techniques the attackers use to create tailored defences, while attackers continuously look for new ways to bypass these defences. It's like a never-ending game of cat and mouse, with each side constantly trying to outsmart the other.

NIST Cyber Security Framework: Defenders use this framework to create a solid foundation for their cybersecurity efforts. During the "Identify" phase, they assess their assets and vulnerabilities, while attackers might use this information to target weak points. In the "Protect" phase, defenders implement safeguards, and attackers look for ways to bypass them. The "Detect," "Respond," and "Recover" phases involve constant monitoring, adapting, and learning for both sides, ensuring they're always ready for the next round of wild west shenanigans.

In a nutshell, each of these frameworks offers unique insights and strategies for both defenders and attackers in the cybersecurity showdown. By understanding and leveraging these frameworks, both sides can better anticipate each other's moves, leading to a more secure (and entertaining) digital landscape. So, saddle up, partner, and let the battle commence!

It's a trap!
Nope, incident response plan!

Ladies and gents, gather 'round the campfire, because I'm about to spin you a yarn about the legendary Incident Response Plan. You see, in the wild world of cybersecurity, threats are like those pesky mosquitoes buzzing around your BBQ – always looking for a way to ruin the fun. And just like you'd have a plan to swat those bugs away, you need a plan to deal with cyber incidents.

Picture this: You're hosting the most epic backyard shindig, complete with finger-lickin' good food, groovy tunes, and good ol' fashioned fun. Suddenly, a swarm of uninvited bees decides to join the party, and chaos ensues. If you'd had an Incident Response Plan, you'd know exactly what to do – grab that trusty bug zapper, call in your beekeeper buddy, and get those stingers out of there!

Incident Response Plans are just like that – they help you prepare for, deal with, and recover from cybersecurity incidents, ensuring your digital shindig goes off without a hitch. So, let's dive into the steps for writing an Incident Response Plan that'll have you feeling like the cybersecurity equivalent of a bee-wrangling superhero.

Step 1: Identify the Hoedown Hiccups

27

Before you can whip up an Incident Response Plan, you've got to know what kind of incidents you might face. You wouldn't plan a barn dance without knowing which critters could crash it, right? So, start by listing all potential cybersecurity incidents, such as malware infections, data breaches, or even those pesky insider threats (like that sneaky cousin who's always trying to swipe your secret pie recipe).

Example: For a company handling sensitive customer data, some incidents to identify might include phishing attacks, ransomware infections, unauthorized access to data, and accidental data leaks.

Step 2: Assemble Your Cyber Posse

You can't wrangle cyber threats on your own – you'll need a team of trusty experts to help you out. Designate roles and responsibilities for each member of your Incident Response Team (IRT). Make sure you've got all the key roles covered, like the incident commander (the sheriff in charge), the security analyst (your trusty deputy), and the communication specialist (the town crier).

Example: Your IRT could include IT staff, security experts, PR representatives, and legal counsel, all ready to jump into action when an incident occurs.

Step 3: Define Your Line Dance (Incident Response Procedures)

Now that you know your potential incidents and have a team at the ready, it's time to lay out the steps to follow when an incident occurs. Think of it like choreographing a line dance – everyone needs to know their moves and when to execute them.

Example: Your response procedures might include steps like detecting the incident, assessing its impact, containing the threat, eradicating the problem, recovering affected systems, and learning from the incident to improve future responses.

Step 4: Communication is Key (Like Yelling "Yeehaw!")

In any hoedown, communication is crucial. Establish guidelines for communication within your team and with external stakeholders, such as customers, partners, and regulators. Make sure your communication is clear, timely, and accurate.

Example: Create templates for incident notifications and updates, designate a spokesperson for external communications, and establish a process for informing relevant parties (such as law enforcement) if needed.

Step 5: Keep Your Boots on the Ground (Practice Makes Perfect)

You wouldn't step onto the dance floor without practicing your moves, would you? The same goes for your Incident Response Plan. Regularly test and update your plan to ensure it remains effective and relevant as new threats emerge and your organization changes.

Example: Schedule periodic tabletop exercises, during which your IRT can walk through simulated incidents and evaluate the effectiveness of their response procedures. This will help identify any gaps or areas for improvement in your plan.

By following these steps and keeping your sense of humour intact, you'll have a rock-solid Incident Response Plan that'll have you two-stepping your way to a more secure cyber environment. Yeehaw!

Tools of the Trade
Navigating the Cyber Security Frameworks Jungle

In this chapter, we ventured into the intricate world of cyber security frameworks. We compared and contrasted three popular frameworks: the Cyber Kill Chain, MITRE ATT&CK, and NIST Cyber Security Framework, each designed to offer guidance and strategies for organizations to protect their digital assets. We examined how both defensive and offensive teams utilize these frameworks to anticipate, prevent, and respond to cyber incidents.

We broke down the key concepts and phases of each framework, providing real-world examples and insights into their practical applications. This chapter equipped readers with valuable knowledge and understanding of these tools, preparing them to face the complexities of the cyber realm in the chapters to come.

Chapter Three: Zerg Rush, The Malware Menace
Battling the Cyber Hordes

Introduction to Malware

Imagine, you're knee-deep in a heated StarkKroft battle, micro-managing your units with precision and cunning, when suddenly, out of nowhere, a swarm of Zerglings floods your base, wreaking havoc on your carefully crafted defenses. Ouch, right?

Well, imagine if those pesky Zerglings were actually insidious bits of malware, creeping their way into your computer systems, ready to wreak havoc on your virtual empire. In this chapter, we'll delve into the dark world of malware, its different forms, and how it can infect and compromise your digital realm, much like the relentless Zerg rush that has crushed the dreams of many a StarkKroft player. Together, we'll strategize and learn the best ways to fortify our defences and neutralize these cyber threats before they have the chance to morph into unstoppable Ultralisks. So, strap in, and let's embark on this journey into the malware-infested depths of cyberspace, armed with knowledge and our love for a good ol' RTS classic.

Malware, short for malicious software, comes in various shapes and forms, each with its own unique way of infiltrating and causing chaos in your digital world. Here are some of the most common types of malware you may encounter:

Virus: Much like their biological counterparts, computer viruses attach themselves to legitimate files or programs and replicate when that file or program is executed. They can corrupt or modify data, cause system crashes, or spread to other devices via email or file sharing.

Worm: Worms are self-replicating pieces of malicious software that spread through networks, exploiting vulnerabilities in operating systems or software. They can consume system resources, cause network slowdowns, and create backdoors for other malware to enter.

Trojan: Named after the famous Trojan horse, this type of malware disguises itself as a legitimate program or file to trick users into downloading and installing it. Once inside your system, Trojans can steal sensitive data, create backdoors for unauthorized access, or even hijack your computer's resources.

Ransomware: This particularly nasty form of malware encrypts your files and holds them hostage, demanding a ransom payment (usually in cryptocurrency) in exchange for the decryption key. Ransomware attacks can cripple businesses and lead to significant financial losses.

Spyware: As the name suggests, spyware is designed to snoop on your digital activities, collecting sensitive information like passwords, credit card numbers, and browsing habits. This data can then be used for identity theft, fraud, or sold to third parties.

Adware: While not always malicious, adware can be annoying and invasive. It displays unwanted ads on your computer, often in the form of pop-ups or banners. Some adware may also track your browsing habits and use that information to serve targeted ads.

Rootkit: Rootkits are stealthy malware that can hide deep within your system, granting unauthorized access and control to hackers. They can be incredibly difficult to detect and remove, often requiring specialized tools and expertise.

Keylogger: Keyloggers record your keystrokes, capturing sensitive information like login credentials and personal data. They can be used for identity theft, corporate espionage, or other nefarious purposes.

Fileless Malware: This type of malware resides in your system's memory rather than on the hard drive, making it difficult to detect and remove using traditional antivirus tools. It can be used for data theft, system compromise, or as a launching point for other attacks.

Remember, understanding these different types of malware is just the first step in building a solid defense against them. Stay vigilant, keep your systems updated, and always practice good cybersecurity hygiene to minimize the risks posed by these digital threats.

Viral Infections in the Digital Realm
When Computer Viruses Go Wild

So, you're at the annual family reunion, and you can't wait to dig into Aunt Karen's famous potato salad. Little do you know, that delightful dish is harboring a nefarious secret: a stomach virus that, like a chain reaction, is about to wreak havoc on everyone's digestive systems. Now, imagine this in the realm of StarkKroft, where the Zerg

33

race is rapidly infesting and multiplying across the galaxy, taking over everything in their path. Computer viruses operate in a similar way, latching onto unsuspecting files or programs, ready to cause digital mayhem.

Just like how Aunt Karen's potato salad spread its "joy" throughout the family or the Zerg's infestation tactics, computer viruses attach themselves to legitimate files or programs and replicate when that file or program is executed. They're the digital equivalent of a sneeze that spreads the flu at a crowded concert or a Zergling's voracious appetite. And while you can't exactly give your computer chicken soup and a warm blanket, you can certainly take measures to protect it from these virtual germs.

For example, imagine you're downloading a free game that promises hours of entertainment—let's say, a fan-made StarkKroft. Unbeknownst to you, this seemingly innocent game is carrying a computer virus. Once installed and executed, the virus quickly spreads to other files and programs, corrupting your precious data, causing system crashes, and clogging your inbox with spam emails that spread the virus even further—like a swarm of Zerglings on a rampage.

Computer viruses can be a real pain in the motherboard, but fear not! Just like washing your hands and getting vaccinated can protect you from biological viruses, or how Terran forces can build defences against Zerg invasions, maintaining up-to-date antivirus software and practicing safe browsing habits can help keep your computer healthy and virus-free. So, the next time you're tempted to download that "too good to be true" free game or that StarkKroft spinoff, remember Aunt Karen's potato salad and the Zerg infestation and think twice.

Digital Worms
The Crawling Chaos of Cyberspace

You're a brave Terran commander in the StarkKroft universe, defending your base against the relentless Zerg onslaught. Suddenly, you notice a Nydus Worm bursting through the ground, spewing out waves of ravenous Zerglings, Hydralisks, and other unsavory critters. That's the same level of chaos a computer worm can unleash in your digital world! But what if I told you that our dear Aunt Karen might be an unwitting accomplice in this intergalactic cyber mayhem?

Worms are self-replicating pieces of malicious software that spread through networks, exploiting vulnerabilities in operating systems or software, much like the Zerg's Nydus Worms create their own network of tunnels to move rapidly between bases. They can consume system resources like a ravenous Zergling pack at an all-you-can-eat buffet, cause network slowdowns as if Banelings were rolling through rush-hour traffic, and create backdoors for other malware to enter—like opening up a secret passage for a Protoss Dark Templar to waltz in and slice your data into confetti.

Imagine our beloved Aunt Karen, who just got her first computer and eagerly forwards chain emails to the entire family. Little does she know, one of those innocent-looking messages contains a worm using her contact list to spread its digital destruction far and wide. It's like

Aunt Karen accidentally built a Nydus Worm right in the middle of our family network!

Historically, worms have caused widespread damage to computer systems around the globe. Remember the infamous "ILOVEYOU" worm from 2000? That little cyber pest caused billions of dollars in damage and spread to millions of computers in just a matter of days! It's as if the Zerg launched a full-scale invasion, and Aunt Karen was the one opening the gates to their insatiable hordes.

So, how do you protect your digital domain from the StarkKroft-style invasion of these pesky worms, and prevent Aunt Karen from accidentally joining forces with the Zerg? Simple: Be proactive and channel your inner Terran commander! Keep your software and operating system up-to-date, like a well-maintained Siege Tank ready for action. Use strong passwords as if they were bunker walls, keeping the Zerg horde at bay. And regularly scan your system for vulnerabilities that these worms could exploit—it's like setting up Sensor Towers and Photon Cannons around your base to detect cloaked enemies.

Finally, don't forget to educate Aunt Karen and the rest of the family on safe email practices, like not opening suspicious attachments or clicking on unknown links. That way, you'll not only save your computer from a Zerg-infested digital apocalypse but also keep your family reunions free of both potato salad-induced stomach bugs and cyber worm pandemonium!

Digital Trojans
When Gifts of Friendship Turn Into Devious Sabotage

Gather 'round, folks, for the tale of the Trojan Horse! No, not the one from ancient mythology, but the digital kind that sneaks into your computer and unleashes all sorts of chaos. Like the cunning scheme from the famous story, Trojan malware comes bearing "gifts" that hide a nasty surprise. Think of it as a seemingly harmless package from Aunt Karen that, once unwrapped, unleashes a horde of angry Zerg into your living room.

Trojans are malicious software masquerading as legitimate applications or files, duping you into installing them on your system. You might think you're downloading the latest StarkKroft mod, but in reality, you're inviting a powerful Dark Archon to wreak havoc on your digital world. Trojans can give cybercriminals remote access to your computer, allowing them to steal your sensitive data, recruit your system into a botnet army, or use it as a launchpad for further attacks on your network.

Now, let's say Aunt Karen, who's recently become a big fan of the Protoss, sends you an email with an attachment that claims to be an exclusive StakKroft wallpaper featuring her favorite Protoss hero, Artanis. You trust Aunt Karen, so you click on the attachment, only to find out that it's actually a Trojan that has infiltrated your system. By the time you realize what's happened, your computer is already under siege, and your data is being spirited away like minerals stolen by a crafty group of Probes.

In the cyber battlefield, Trojans are the ultimate double agents, cunningly disguised as something innocuous while harboring a hidden agenda. As a defender against these digital menaces, you must be ever-vigilant, questioning even the most seemingly harmless files and applications. Keep your antivirus software up to date, like a vigilant Dragoon patrolling your base's perimeter. Be cautious when downloading files or clicking on links, especially if they come from unknown sources.

Remember, while Aunt Karen might have good intentions, she may unwittingly be a pawn in a larger cyber scheme. Educate her and your friends about the dangers of Trojans and the importance of verifying the safety of files and links before sharing or downloading them. With a keen eye and a healthy dose of caution, you can keep these digital Trojan Horses from breaching the walls of your cyber fortress and turning your computer into a battleground between the Terrans, Zerg, and Protoss.

Digital Hostage-Taking
Ransomware and the Ultimate Cyber Shakedown

Now, you're hosting the biggest StarCraft tournament of the year, with Aunt Karen leading the charge as the most enthusiastic participant. Everything is going swimmingly until suddenly, the screen goes black, and a sinister message appears, demanding payment in exchange for the return of your precious game data. That's ransomware for you—the digital equivalent of having your prize-winning Zerg army held hostage by a rogue group of Protoss pirates.

Ransomware is a type of malware that encrypts your files and data, rendering them inaccessible, and then demands a ransom (usually in cryptocurrency) for the decryption key. It's as if a cunning Infestor has cast a Fungal Growth on your computer, immobilizing your digital assets and holding them hostage until you cough up the cash.

You might encounter ransomware through a phishing email, like one from "Aunt Karen" that contains a link to a website claiming to have the latest StarCraft expansion pack (spoiler alert: it doesn't!). Or you may stumble upon it while downloading a seemingly innocuous software update. One wrong click, and suddenly your files are locked up tighter than a maxed-out Zerg base swarming with Spore Crawlers.

So how do you protect your digital empire from these ransom-hungry cyber bandits? Here's the strategy: First, regularly back up your data like you're stockpiling Vespene Gas for a massive Battlecruiser fleet. That way, even if ransomware strikes, you'll have a backup to fall back on, and you won't be at the mercy of these digital extortionists. Second, keep your antivirus software and system updates current, like a well-oiled Siege Tank ready to blast any intruders. And finally, exercise caution when opening emails, clicking on links, or downloading files, especially from unknown sources. Don't let a tempting "Aunt Karen" offer lure you into a ransomware trap.

Adopting these defensive tactics can minimize the risk of ransomware taking your files hostage and derailing your epic StarCraft tournament. Remember, knowledge is power in the battle against cyber threats, and vigilance is key. Now go forth and conquer the digital realm, one Zergling rush at a time!

Digital Espionage
Spyware and their Sneaky Sabotage

You're amidst an intense StarCraft match, executing a top-secret strategy that will surely lead you to victory. Suddenly, you notice that your opponent seems to know your every move, as if they've tapped into your mind and are watching your every click. That's the digital deceit of spyware—a cunning observer hiding in the shadows, collecting your personal information and sending it back to its creators.

Spyware is a type of malware that stealthily monitors your computer activity, collecting sensitive information like passwords, financial data, and even your game strategies (that's right, Aunt Karen, we're onto you!). It can be unknowingly installed on your computer through deceptive downloads or phishing emails—think of it as a Protoss Observer cloaked in the shadows, waiting to gather intel on your most confidential plans.

Perhaps you receive an email from "Aunt Karen" with an attachment claiming to contain top-secret tips for mastering your StarCraft gameplay. You eagerly download the file, not realizing you've just invited a digital spy into your home. Now, every keystroke you make and every move you execute is being monitored and recorded, all while your opponent (Aunt Karen, of course) uses this intel to dominate the battlefield.

So, how do you protect your computer and your carefully crafted StarCraft strategies from the prying eyes of spyware? First, invest in a robust antivirus and anti-spyware program that scans for and removes these digital intruders, like a vigilant Terran Ghost hunting down cloaked enemies. Second, keep your software and operating system updated, as if you're researching the latest upgrades at your Engineering Bay. And finally, be cautious when opening emails or downloading files, especially from unknown sources. Remember, even an email from "Aunt Karen" could be harboring a hidden spy!

You can mitigate the risk through good policies, staying vigilant and taking the necessary precautions, you can keep spyware from infiltrating your digital domain and stealing your StarCraft secrets. With your cyber defences in place, you'll be well on your way to claiming victory in the Koprulu Sector and beyond!

Adware Annoyances
When Your Digital Battlefield is Bombarded with Ads

Now you're deep in a strategic StarCraft battle, carefully managing your resources and building an impressive army. Suddenly, your screen is bombarded with pop-up ads for "Aunt Karen's Terran-Approved Kitchen Gadgets" or "Zerg-Proof Hair Conditioner." The constant barrage of distractions is frustrating and wreaking havoc on your focus. Welcome to the world of adware, where the enemy is irritating advertisements and a relentless Aunt Karen.

Adware is a type of malware that displays unwanted advertisements on your computer, often in the form of pop-ups, banners, or even redirects. While not always malicious, adware can be annoying and invasive, disrupting your StarCraft battles or slowing down your system like a lumbering Ultralisk trudging through molasses. And, of course, there's always the risk that the ads themselves could lead you to more dangerous malware or scams.

Picture this: you're browsing the web, searching for tips on how to improve your StarCraft game. You stumble upon a website offering a free download of "Aunt Karen's Ultimate StarCraft Strategy Guide." But when you download the guide, you're not just getting tips and tricks—instead, you've inadvertently invited adware into your computer. Now, every time you play StarCraft or surf the web, you're bombarded with ads for Aunt Karen's various products, schemes, and pyramid-shaped investment opportunities.

So, how do you defend your digital domain from adware and keep your StarCraft battles free from Aunt Karen's pesky ads? First, make sure you have a reliable antivirus program that also scans for and removes adware, like a vigilant Marine squad clearing out Zerg Creep Tumors. Second, keep your software and operating system up-to-date, as if you're constantly refining your tech tree. And lastly, be cautious when downloading files or clicking on links, especially from unfamiliar sources. Not every "free" strategy guide is worth the price of admission.

By taking these precautions, you can keep your computer free from adware and maintain a clear, focused battlefield for your StarCraft matches. Aunt Karen may be relentless, but with your defenses in place, her ads won't stand a chance!

Rootkits
The Dark Archons of Malware, Where Aunt Karen's Deception Runs Deep

In the StarCraft universe, the Dark Archon is a powerful and deceptive Protoss unit capable of manipulating the minds of its enemies and turning them against each other. Much like the Dark Archon, a rootkit is a type of malware that hides in plain sight, lurking deep within your computer's system and using its stealth to control and manipulate your device from the shadows. And, as always, Aunt Karen's scheming ways can never be underestimated.

Rootkits are insidious pieces of software that can grant an attacker control over your computer or network without your knowledge. They can hide their presence by cloaking themselves within legitimate processes, making them difficult to detect and remove. It's as if Aunt Karen had secretly replaced your loyal Dragoons with mind-controlled enemy units, waiting to strike when you least expect it.

Imagine you download a program that promises to improve your StarCraft gameplay, only to find out later that it was actually a rootkit in disguise. Now, Aunt Karen has access to your computer, able to monitor your every move, steal sensitive information, and even use your device as a launchpad for further attacks. Worst of all, you might not even realize that she's watching your every click and keystroke.

So, how do you defend your digital domain from the Dark Archon-like deception of rootkits? First, ensure you have a robust security suite with specialized rootkit detection and removal capabilities, like a squad of Observers keeping a watchful eye on your base. Second, keep your software and operating system up-to-date, like upgrading your Protoss technology to stay ahead of the enemy. And lastly, be cautious when downloading files or clicking on links, especially from unfamiliar sources— some "helpful" programs may actually be Trojan horses for rootkits.

You can protect your computer from the stealthy and dangerous world of rootkits by implementing best practices and policies. This ensures that Aunt Karen's deception doesn't find a foothold in your digital realm. With your defences in place, not even the craftiest Dark Archon—or Aunt Karen—will be able to penetrate your system!

Keyloggers
The Ghosts of Malware That Record Your Every Move

Imagine you're a Terran Ghost in the StarCraft universe, an elite covert operative trained to gather intelligence and infiltrate enemy lines without being detected. You're the ultimate spy, capable of capturing critical information and turning the tide of battle. Now, picture Aunt Karen employing a similar tactic in the digital realm, using keyloggers as her sneaky spies to snoop on your every move.

Keyloggers are a type of malware designed to record and transmit every keystroke you make on your computer.

These digital spies can be stealthy, hiding in the background as you type in passwords, credit card numbers, and other sensitive information while reporting back to Aunt Karen—or any other cybercriminal who's deployed the keylogger. It's like having a Ghost cloaked and hovering over your shoulder, noting down your every move.

Let's say you've installed a new voice chat app to communicate with your fellow StarCraft players. Unbeknownst to you, that app came bundled with a keylogger that's now silently monitoring your every keystroke. As you type in your account credentials and other personal information, the keylogger is transmitting that data straight to Aunt Karen's secret base, where she can use it for all manner of nefarious purposes.

So how can you defend against these digital Ghosts and keep Aunt Karen's prying eyes at bay? First, invest in a reliable security suite with anti-keylogger capabilities, like a group of Science Vessels equipped with EMP Shockwaves to neutralize enemy cloaking. Second, keep your software and operating system updated to patch any vulnerabilities that keyloggers might exploit, like bolstering your defenses with missile turrets and sensor towers. Lastly, be cautious when installing new programs, predominantly from unknown sources—some seemingly innocuous software might actually be a cover for keylogger infiltration.

By staying alert and taking these precautions, you can ensure that your keystrokes remain private and secure, preventing Aunt Karen's keylogging spies from gaining access to your sensitive information. So, the next time you sit down at your computer to strategize your

StarCraft victory, you can do so confidently, knowing that your digital realm is safe from the prying eyes of keyloggers and Aunt Karen's sneaky tactics.

Fileless Malware
The Protoss Dark Templar of Cyberthreats

In the StarCraft universe, the Protoss Dark Templar are elite stealth warriors, able to slip past enemy defenses undetected thanks to their advanced cloaking abilities. They strike without warning, leaving chaos in their wake before vanishing into the shadows. Now, imagine if Aunt Karen were to harness the power of these elusive warriors and unleash them in the digital realm. Enter fileless malware—the Dark Templar of cyber threats.

Fileless malware is a unique breed of malicious software that operates entirely in your computer's memory, leaving no trace on your hard drive. It's like a Dark Templar infiltrating your base, phasing in and out of existence while wreaking havoc on your precious data. Traditional antivirus software struggles to detect these elusive threats since they don't leave any files behind for scanning, making them an ideal weapon for Aunt Karen and her cybercriminal cohorts.

Picture this: You're browsing your favourite StarCraft forum when you stumble upon an intriguing link promising exclusive tips on how to counter a Zerg rush. Excited, you click on it, only to unwittingly trigger a fileless malware attack. The malware exploits a vulnerability in your browser, injecting malicious code directly into your computer's memory. Just like a Dark Templar slipping past your defenses, the malware starts

46

to gather sensitive information and manipulate your system—all while remaining virtually invisible to your antivirus software.

So, how do you protect your digital fortress from Aunt Karen's invisible menace? First, keep your operating system and software updated, like a vigilant Protoss Sentry casting Guardian Shield to protect its allies. This will help patch any vulnerabilities that fileless malware could exploit. Second, invest in a robust security suite with behavioral monitoring and memory scanning capabilities, like a well-positioned Photon Cannon ready to reveal cloaked threats. Finally, practice safe browsing habits, avoiding suspicious links and downloads that could open the door for fileless malware attacks.

By following these tips, you can keep your computer safe from the Dark Templar of cyber threats and thwart Aunt Karen's fileless malware machinations. So go forth, fearless StarCraft commander, and conquer the galaxy knowing that your digital realm is well-guarded against the invisible menace of fileless malware.

The Menagerie of Malware
The crimes of the Grindelworld

In this thrilling StarCraft-inspired adventure, we delved deep into the treacherous world of malware, exploring the various types of digital threats that lurk in the shadows of cyberspace. From viruses that spread like Aunt Karen's infamous potato salad to worms that wreak havoc like Zerglings in an all-you-can-eat buffet, we've dissected the strategies and tactics these cyber-nasties employ to invade our digital lives.

47

We also tackled the elusive fileless malware, the Dark Templar of cyber threats, and the cunning Trojans, which disguise themselves as innocent programs. Furthermore, we learned about ransomware's devastating hostage-taking tactics, the sneaky spyware that watches our every move, the annoying adware that plagues our browsing experience, the powerful rootkits that dig deep into our systems, and the stealthy keyloggers that record our every keystroke.

Armed with this newfound knowledge, we can channel our inner StarCraft commander to protect our digital fortresses from Aunt Karen's cyber minions. By keeping our systems updated, using robust security software, and practicing safe browsing habits, we can conquer the galaxy of malware threats and safeguard our digital realm from the insidious forces that seek to exploit it.

In the upcoming Chapter 4, we will dive into the initial reactions to a cyber security incident. Much like a surprise Zerg attack on your well-fortified base, these incidents can catch even the most seasoned commanders off guard. But fear not, because we'll explore how to handle the critical first 48 hours of an incident, just like a true StarCraft hero would.

Chapter Four: Invasion Alert, The First 48 Hours
Battle Stations, Commanders!

Introduction to Initial Reactions

It's a normal day at the office, everyone's busy working (or browsing Facebook), and then suddenly, your computer screen flashes with a terrifying message: "Your data has been encrypted. Pay up, or say goodbye to your precious files!" Yup, you've just experienced a cyber security incident, and it's about to turn your workplace upside down.

In this chapter, we'll take you through the instinctual reactions to cyber incidents, with all the chaos, confusion, and bizarre behaviour you'd expect from an episode of "The Office" (spoiler alert: Michael Scott is not the ideal crisis manager). We'll explore the initial moments of panic and the essential steps to regain control of the situation and minimize the damage.

Throughout the chapter, we'll cover topics such as:

- The five stages of cyber incident grief: denial, anger, bargaining, depression, and acceptance (all while maintaining a sense of humour, of course).

- Real-world examples of cyber incidents and how organizations have handled them—both the good, the bad, and the downright hilarious.

- The importance of staying calm and focused in the face of cyber adversity (channelling your inner Jim Halpert, perhaps?).

- Best practices for communication and teamwork during a cyber incident, with a special emphasis on avoiding a Dwight Schrute-style office coup.

- Strategies for assessing and triaging the situation and prioritizing your response efforts like a true cyber security pro.

So, take a deep breath, grab your "World's Best Boss" mug, and join us as we delve into the wild world of initial reactions to cyber security incidents. Remember, when it comes to cyber incidents, staying calm and working together can make all the difference. And if all else fails, there's always a "Threat Level Midnight" movie night to look forward to!

The Five Stages of Cyber Incident Grief
When Kubler-Ross Meets the IT Department

As a cyber incident unfolds, emotions run high and people tend to experience a rollercoaster of feelings, much like the five stages of grief as defined by Elizabeth Kubler-Ross. Strap in, folks, because we're about to explore this emotional journey through a cyber-incident lens!

Denial: "This can't be happening! My computer must be playing a prank on me."

When a cyber incident first strikes, denial is often the knee-jerk reaction. You might think your system is just glitching or that your coworker is playing a cruel joke (looking at you, Jim!). Remember that time in 2013 when the New York Times' website went down, and everyone initially thought it was just a technical issue? Surprise! It was a cyber attack.

Anger: "Who's responsible for this? I want to unleash a thousand pop-up ads on their computer!"

As reality sinks in, anger takes over. Frustration builds as you grapple with the fact that someone out there is messing with your data. It's like when your favourite character in a TV show gets killed off, and you can't help but feel personally attacked (we're still not over you, Ned Stark).

Bargaining: "If only I'd updated my antivirus software, this wouldn't have happened. I promise I'll be more diligent from now on!"

During the bargaining stage, you'll start making deals with the universe, pledging to be a model cyber citizen if only you can undo the damage. It's like when you're trying to recover an unsaved Word document, desperately pleading with your computer to bring back those three hours of work.

Depression: "All our data is gone, our customers are furious, and I just spilled coffee on my keyboard. This is the end."

When the weight of the cyber incident truly hits you, depression can set in. It's like that moment when you're stuck in traffic, late for an important meeting, and "Everybody Hurts" by R.E.M. starts playing on the radio. But don't lose hope! This, too shall pass.

Acceptance: "Okay, this happened. Let's learn from it and become stronger."

Finally, you'll reach the stage of acceptance. This is when you put on your problem-solving hat and tackle the incident head-on, learning valuable lessons and emerging stronger (and hopefully wiser) for the future. Just like when Sony Pictures Entertainment faced a massive cyber attack in 2014, they eventually recovered and used the experience to bolster their cyber security measures.

So there you have it: the five stages of cyber incident grief. By recognizing these emotions and navigating them with humor, you can better cope with the chaos and pave the way for a more resilient response to cyber incidents. Remember, laughter is the best medicine—even in the face of cyber adversity.

Tales from the Cyber Trenches
When Organizations Face Cyber Incidents

Cyber incidents are like those surprise quizzes in school—you never know when they'll strike, and they can leave you feeling flustered and unprepared. In this section, we'll take a stroll down memory lane to explore some real-world examples of cyber incidents and how organizations have handled them. Spoiler alert: not all heroes wear capes; some wield keyboards and a wicked sense of humor!

The Wannabe Ransomware: In 2017, the infamous WannaCry ransomware attack held computers across the globe hostage. The UK's National Health Service (NHS) was particularly hard hit, with surgeries and appointments cancelled left and right. But like an IT superhero, a young cybersecurity researcher (let's call him "Captain Patch") stumbled upon a "kill switch" for the ransomware, saving countless systems from doom. Talk about being in the right place at the right time!

The Great Twitter Hack of 2020: When high-profile Twitter accounts like Elon Musk, Joe Biden, and Bill Gates started tweeting about Bitcoin giveaways, it was clear that something was amiss. A hacker had gained access to these accounts and was having a field day. The response? Twitter's security team went into overdrive, locking down affected accounts and investigating the breach. The moral of the story? Always keep your passwords safer than your secret family recipe for Aunt Karen's potato salad.

The Target Breach: In 2013, retail giant Target suffered a massive data breach, affecting over 40 million customers. While this cyber incident was no laughing matter, Target took it as an opportunity to learn and improve. They didn't just sweep the issue under the rug; they launched a full-scale investigation, offered free credit monitoring for customers, and beefed up their cybersecurity measures. Target's response shows that it's not about how hard you get hit, but how quickly you can bounce back.

The Hilarious Cyber Vigilantes: When scammers tried to con a cybersecurity researcher into revealing his personal information, they had no idea whom they were messing with. The researcher turned the tables on the scammers, hacking their CCTV systems and live streaming their activities on YouTube. The lesson here? Don't mess with cybersecurity experts—they're not your average keyboard warriors!

So, there you have it, folks: a collection of cyber incidents that range from heroic to hilarious. By learning from these examples, we can better prepare ourselves for the inevitable cyber surprises that lie ahead. After all, knowledge is power, and laughter is the best antivirus!

Keeping Your Cool in the Cyber Hot Seat
How to Channel Your Inner Jim Halpert

When you're dealing with a cyber incident, it's easy to let your emotions get the best of you. But just like our favourite paper salesman and prankster extraordinaire, Jim Halpert from The Office, staying calm and focused in the face of adversity can be a game-changer. Let's explore why keeping your cool is crucial when dealing with cyber crises and how you can harness your inner Jim Halpert to ace your response.

The Power of a Level Head: Remember the time when a bat infiltrated the Scranton branch of Dunder Mifflin? Amidst the chaos and panic, it was Jim who maintained his composure and turned the situation into a hilarious prank on Dwight. In the realm of cyber incidents, keeping a level head can help you make better decisions,

54

avoid knee-jerk reactions, and ultimately resolve the issue more effectively.

The Great Sony Hack: In 2014, Sony Pictures was hit by a devastating cyber attack that leaked confidential data and crippled their systems. The company's executives could have easily lost their cool, but they worked methodically to address the situation instead. They communicated with employees, collaborated with law enforcement, and even released the movie at the center of the controversy. That's the power of staying calm under pressure!

The OPM Data Breach: When the U.S. Office of Personnel Management (OPM) discovered a massive data breach in 2015, it was undoubtedly high-stress. But instead of panicking, the OPM leadership focused on remediation efforts, working with other agencies to mitigate the damage and prevent further breaches. It's like when Jim organized the Office Olympics during a tense day at Dunder Mifflin—sometimes, a little composure can make all the difference.

The Zen of Incident Response: Channeling your inner Jim Halpert during a cyber crisis means keeping your wits about you and not getting overwhelmed by the chaos. Take a deep breath, gather your team, and tackle the issue step by step. And hey, if you can slip in a well-timed prank or two (metaphorically speaking) to lighten the mood, all the better!

In conclusion, staying calm and focused in the face of cyber adversity is an invaluable skill. Embrace your inner Jim Halpert, and you'll be better equipped to handle whatever cyber surprises come your way. And remember, when in doubt, ask yourself, "What would Jim do?"

Navigating the Cyber Incident Maze
Communication and Teamwork Tips

Ah, Dwight Schrute—the bumbling, overzealous, beet-loving Assistant (to the) Regional Manager from The Office. While his antics and misguided ambitions can be entertaining, you definitely don't want a Dwight-style coup during a cyber incident. So, let's talk about best practices for communication and teamwork when facing cyber threats, all while keeping the spirit of Dunder Mifflin alive.

Clear Communication is Key: Just as Michael Scott's infamous "Conference Room, FIVE MINUTES!" announcements brought the team together, timely and clear communication is crucial during a cyber incident. Ensure everyone is on the same page, understands their roles, and knows what's expected of them. Avoid Dwight-style cryptic messages or unnecessary panic.

The Power of Collaboration: Remember when Jim, Pam, and the rest of the office banded together to prank Dwight? That's the kind of teamwork you need during a cyber crisis. Foster a collaborative environment where everyone feels empowered to contribute, ask questions, and share ideas. You'll be amazed at the creative solutions emerging when everyone works together.

Lessons from the Wannacry Ransomware Attack: In 2017, the WannaCry ransomware attack hit organizations worldwide, causing massive disruptions. But a united effort from cybersecurity professionals, law enforcement, and companies helped mitigate the damage and eventually stop the attack. Much like when Dunder Mifflin's employees joined forces to save the Michael Scott Paper Company, teamwork makes the cyber-dream work!

The No-Dwight Zone: Dwight Schrute's insubordinate attitude and attempts at seizing power were the stuff of comedy legend. However , that kind of behaviour can be disastrous in a cyber incident. Encourage a culture of respect, cooperation, and open communication to ensure everyone stays focused on the task at hand and not on office politics.

Celebrate Success Together: Once the cyber incident has been resolved, don't forget to acknowledge and appreciate your team's hard work. Remember when Michael held the Dundie Awards to celebrate his employees' achievements (and quirks)? While you may not need a tacky trophy ceremony, showing gratitude and recognizing everyone's contributions can go a long way in boosting morale and fostering a positive work environment.

Effective communication and teamwork can be the difference between a successful cyber incident response and a Dwight Schrute-esque disaster. Keep these best practices in mind, and you'll be ready to navigate any cyber challenge that comes your way, all while maintaining that Dunder Mifflin charm.

Cyber Incident Triage
Channelling Your Inner Cybersecurity Pro to Prioritize and Conquer

In the middle of a cyber crisis, it can be easy to feel like you're being swarmed by Zerglings from all sides. But fear not, intrepid cyber defenders! With the right strategies in place, you can assess, triage, and prioritize your response efforts like a true cyber security pro (or a seasoned StarCraft player).

Keep Calm and Triage On: Remember when Michael Scott accidentally burned his foot on a George Foreman Grill? The chaos that ensued was less than ideal. To avoid a similar meltdown during a cyber incident, stay calm and focused. Gather information, assess the situation, and determine the severity of the threat before taking any action.

Prioritize Like a Boss: Not all cyber threats are created equal. Some are like Dwight Schrute's fire drill fiasco, while others are more like Jim Halpert's harmless pranks. Determine which threats require immediate attention and which can be dealt with later. Prioritizing allows you to allocate resources efficiently and minimize potential damage.

Lessons from the Sony Pictures Hack: In 2014, Sony Pictures Entertainment fell victim to a massive cyber attack that exposed sensitive data and disrupted operations. The company's response was criticized for being slow and disorganized. Learn from their mistakes by having a clear plan in place for assessing and triaging cyber incidents, so you can spring into action like a Protoss Zealot when the need arises.

The Incident Commander Role: Assign an incident commander to oversee and coordinate the response effort. This person should be like the wise and collected Captain Picard of the cyber world—able to calmly assess the situation, prioritize tasks, and lead their team through the chaos.

Test, Test, Test: Much like practicing your StarCraft build orders and strategies, regularly test your incident response plan. Run drills and simulations to ensure everyone knows their roles and responsibilities, and can effectively assess, triage, and prioritize during a real cyber incident.

In a nutshell, assessing and triaging a cyber incident is all about staying cool under pressure, prioritizing threats, and working together as a team. With these strategies in your cyber arsenal, you'll be ready to tackle any cyber crisis that comes your way, all while maintaining that laid-back, Office-inspired charm.

Fight or Flight
Play dead, or run in dread!

This chapter took us on a wild ride through the world of initial reactions to cyber incidents. We laughed, we cried, and we learned a thing or two about the five stages of cyber incident grief, real-world examples of how organizations have handled incidents, the importance of staying calm and focused (à la Jim Halpert), best practices for communication and teamwork (avoiding Dwight-style office coups), and strategies for assessing and triaging situations like a true cyber security pro.

With all that under our belts, it's time to transition into the next phase of our cyber journey: Chapter 5. Here, we'll explore the ins and outs of crafting a comprehensive, actionable, and—dare we say—fun IRP that will guide your organization through cyber incidents with grace, efficiency and a touch of that trademark humour we've come to love. So please put on your helmets, cyber enthusiasts, and let's dive into the exciting world of Incident Response Plan (IRP) creation!

Chapter 5: Crafting Your Incident Response Plan (IRP)

Defending Your Cyber Space from Alien Invaders

Introduction of Writing an IRP

When it comes to dealing with cyber incidents, having a solid Incident Response Plan (IRP) in place is as essential as Will Smith and Jeff Goldblum teaming up to save humanity from an alien invasion. It's time to channel our inner President Whitmore, roll up our sleeves, and craft an IRP that will make even the most determined cyber attacker think twice about messing with our digital domain.

An incident response plan can contain anywhere between four to nine stages.

Although, deviating from the standard in this book. I maintain the following as a reference:

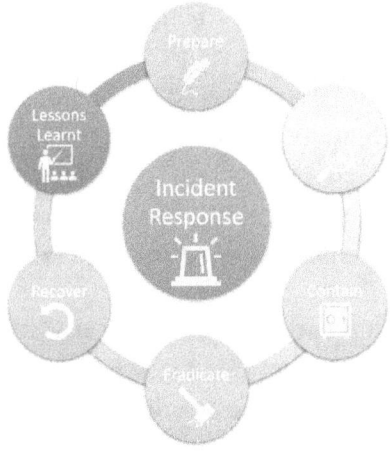

In this action-packed chapter, we'll explore the following:

- The core components of an effective IRP: Just like how the movie 'Independence Day' had an ensemble cast of heroes, an IRP has its own set of essential elements that work together to create a winning strategy. We'll delve into the critical components of an IRP, from assembling your dream team of cyber defenders to crafting a communication plan that rivals President Whitmore's iconic speech.

- Preparing for the unexpected: Let's face it, not even the characters in 'Independence Day' could have predicted the extent of the alien invasion. Similarly, when it comes to cyber incidents, the unexpected is often the norm. We'll discuss how to prepare for various types of incidents, assess potential risks, and ensure your organization is ready to tackle anything that comes its way.

- A step-by-step guide to creating your IRP: With the fate of the digital world in our hands, it's time to put pen to paper (or fingers to keyboard) and craft an IRP that would make even Area 51 proud. We'll walk you through each process stage, providing real-world examples and helpful tips to ensure your IRP is as comprehensive and effective as possible.

- Testing and updating your IRP: An IRP is a living document that requires regular maintenance, much like keeping those pesky alien spaceships in check. We'll explore best practices for testing and updating your IRP, ensuring it remains relevant and ready to handle the ever-evolving cyber threat landscape.

- Learning from the (fictional) experts: We'll wrap up the chapter by drawing inspiration from the heroes of 'Independence Day,' showcasing how their resourcefulness, adaptability, and teamwork can serve as valuable lessons for anyone tasked with defending their organization from cyber threats.

As we reach the end of our thrilling journey through the world of Incident Response Plans, we can confidently say that we're ready to tackle any cyber invader that dares to threaten our digital domain. With an effective, well-crafted IRP in hand, we can boldly declare, "Today, we celebrate our Independence Day!"—or, at the very least, our triumph over the forces of cyber chaos. So, let's raise a glass to our newfound cyber resilience, and may our digital defences always stand firm!

Assembling Your Cyber-Avengers
The Core Components of an IRP Worthy of a Hollywood Blockbuster

When it comes to crafting an effective Incident Response Plan (IRP), you'll want to assemble a team that's just as impressive as the star-studded cast of 'Independence Day.' After all, who wouldn't want to have Will Smith and Jeff Goldblum on their side when facing a cyber invasion? But since they're busy saving the world from extraterrestrials, we'll have to settle for a different kind of dream team—a team of cyber defenders! Here are the core components of an IRP that will help you strategize and protect your digital domain:

Roles and responsibilities: Just as President Whitmore had to rally the troops, your IRP needs a clear definition of roles and responsibilities for your team members. From the cybersecurity analysts identifying threats to the PR team handling communications, everyone should know their role in the event of a cyber incident. This way, you can ensure smooth and efficient coordination, even when the fate of your digital world hangs in the balance.

Communication plan: Remember President Whitmore's spine-tingling speech before the final battle against the aliens? Your IRP needs a communication plan that's just as effective. Outline how your organization will share information internally and externally during a cyber incident. This includes notifying stakeholders, employees, customers, and even the media if necessary. Having a well-structured communication plan will help prevent confusion, misinformation, and panic—leaving you free to focus on defeating the cyber invaders.

Incident detection and analysis: Much like how the heroes of 'Independence Day' had to study the alien invaders to devise a plan of attack, your IRP should detail how your organization will detect and analyze potential cyber incidents. This involves establishing processes for monitoring networks, identifying vulnerabilities, and assessing the severity of incidents. By staying informed about the enemy (aka the hackers), you'll be better equipped to protect your digital turf.

Containment, eradication, and recovery: When it comes to fending off cyber threats, you'll want to channel your inner Captain Steven Hiller (Will Smith's character) and take swift, decisive action. Your IRP should outline strategies for containing and eradicating threats and recovering from any damage caused by the incident. Just like the heroes in 'Independence Day' had to rebuild their world after the alien invasion, your organization must be prepared to restore normal operations and repair any damage resulting from a cyber incident.

Post-incident review: After the dust has settled, it's crucial to learn from the experience and improve your cyber defences, much like how humanity in 'Independence Day' learned valuable lessons from their encounter with the aliens. Conduct a thorough post-incident review to analyze what went well, what could have gone better, and how to prevent similar incidents in the future.

By incorporating these core components into your Incident Response Plan, you'll be ready to face any cyber threat with confidence, just like the brave characters of 'Independence Day'. So, let's get to work, Cyber-Avengers—it's time to defend our digital world from the forces of cyber chaos!

Expecting the Unexpected
Bracing for Cyber Surprises Like a Sci-Fi Hero Facing an Alien Invasion

In the world of cyber incidents, surprises are as common as the plot twists in your favorite sci-fi blockbuster. Just as the characters in 'Independence Day' were caught off guard by the massive alien invasion, your organization might face a cyber threat that's beyond your wildest imagination. But fear not, my digital warriors! With a little bit of foresight and a whole lot of preparation, you can be ready to tackle any cyber surprise that comes your way.

Risk assessment: Just as Jeff Goldblum's character, David Levinson, analyzed the aliens' attack patterns to predict their next move, your organization should conduct regular risk assessments to identify potential threats and vulnerabilities. By understanding your organization's unique risk profile, you can prioritize your resources and develop a robust IRP tailored to your specific needs.

Scenario planning: Remember the chaos that ensued when the aliens first arrived in 'Independence Day'? You can avoid that kind of pandemonium by creating response scenarios for various types of cyber incidents. Think of it as digital disaster preparedness drills: By simulating different cyber threats, your team will be better equipped to handle real-world incidents with confidence and agility.

Training and awareness: Much like how the movie's heroes had to learn about the aliens' technology to stand a chance against them, your employees should be well-versed in cybersecurity best practices. Conduct regular training sessions and awareness campaigns to ensure your team is up to date on the latest threats, vulnerabilities, and incident response strategies.

Building resilience: In 'Independence Day', humanity's survival hinged on their ability to adapt and bounce back from the alien onslaught. Similarly, your organization's resilience is crucial in the face of cyber incidents. Implement robust backup and recovery systems, continuously update your security measures, and foster a culture of preparedness to ensure your organization can withstand and recover from even the most unexpected cyber attacks.

Collaborating with external partners: When battling an alien invasion—or a cyber incident—having allies can make all the difference. Establish relationships with external partners, such as law enforcement, cybersecurity experts, and industry peers, to share threat intelligence, best practices, and support during a crisis.

By embracing the unpredictable nature of cyber incidents and preparing for the unexpected, you'll be ready to face any cyber challenge, just like the heroes of 'Independence Day'. So, buckle up and get ready for a wild ride through the digital universe, as we boldly go where no organization has gone before!

From Area 51 to Your Organization *Crafting an IRP That's Out of This World*

Grab your tinfoil hats, folks! It's time to venture deep into the realm of IRP creation. With the survival of your digital empire at stake, we'll guide you through each step of the process, offering examples and insights to ensure your IRP is as robust and comprehensive as possible. So let's get started and create an IRP that even the top-secret masterminds at Area 51 would envy!

Set your objectives: Before we dive into the nitty-gritty of IRP creation, let's establish your goals. Your IRP should outline your organization's overall approach to incident response, detailing the steps your team will take to identify, contain, eradicate, and recover from cyber

67

incidents. Keep these objectives in mind as you develop your plan, and make sure they align with your organization's broader cybersecurity strategy.

Define roles and responsibilities: Just as the characters in 'Independence Day' each had their unique skills and roles to play, your IRP should outline the roles and responsibilities of everyone involved in the incident response process. This includes your in-house cybersecurity team, external partners, and key decision-makers. Clearly defining these roles will help ensure a smooth and coordinated response when incidents occur.

Develop your response procedures: With your objectives and team in place, it's time to create a playbook of response procedures for various incident types. These procedures should be clear, concise, and actionable, guiding your team through each phase of the incident response process. Be sure to include contingencies for various scenarios, such as data breaches, ransomware attacks, or alien invasions (okay, maybe not that last one).

Create communication protocols: In times of crisis, clear communication is critical. Develop a communication plan that outlines how your organization will share information internally and externally during an incident. This plan should detail the channels and processes for sharing information, as well as guidelines for maintaining confidentiality and preserving evidence.

Establish a post-incident review process: After you've successfully thwarted a cyber attack (or fended off extraterrestrial invaders), it's essential to review and learn from the experience. Establish a process for conducting post-incident reviews, analyzing what went well, and identifying areas for improvement. This feedback loop will help refine your IRP and ensure your organization is continually improving its incident response capabilities.

By following these steps and channeling your inner 'Independence Day' hero, you'll craft an IRP that's out of this world. So let's raise our keyboards and take the fight to the cyber invaders, ensuring our digital realm remains safe and secure for generations to come!

Keeping Your IRP Shipshape
An Ongoing Battle Against Cyber Invaders, Independence Day-Style

You've assembled your cybersecurity dream team, crafted a stellar IRP, and saved the digital world from a colossal cyber meltdown. But don't kick back and relax just yet! Much like those pesky aliens in 'Independence Day' that just wouldn't stay down, cyber threats are an ever-evolving menace that requires constant vigilance. In this section, we'll explore how to keep your IRP up-to-date and battle-ready through regular testing and updates.

Conduct regular exercises: Remember how the heroes of 'Independence Day' honed their skills through intense training and practice? Your organization should also conduct regular exercises to test your IRP and identify any potential gaps or weaknesses. This could involve tabletop exercises, simulations, or even full-scale mock incidents. The more you train, the better prepared you'll be when the real cyber invaders come calling!

Learn from real-world incidents: Just as the movie's characters adapted their tactics based on real-world encounters with the aliens, your organization should learn from actual cyber incidents—both those that impact your organization and those affecting others in your industry. By staying informed and learning from others' experiences, you can continuously refine and strengthen your IRP.

Update your IRP to reflect changes in technology and threats: As we all know, the digital realm is constantly evolving, with new threats emerging and old ones adapting to new tactics. It's essential to keep your IRP current by incorporating information about the latest threats, vulnerabilities, and defensive measures. Think of it as upgrading your arsenal of alien-fighting weapons to stay one step ahead of the enemy!

Keep your team informed: Just as 'Independence Day' saw various groups joining forces to combat the alien invasion, it's vital that your entire organization stays informed about changes to your IRP. Regularly communicate updates, lessons learned, and best practices to ensure everyone is on the same page and ready to spring into action when needed.

Schedule periodic IRP reviews: An IRP is a living document, and like any good sci-fi sequel, it needs to evolve to stay relevant. Schedule regular reviews of your IRP, ideally involving key stakeholders from across your organization, to ensure it remains effective and up-to-date in the face of an ever-changing cyber landscape.

By following these best practices for testing and updating your IRP, you'll ensure your organization remains prepared and agile in the face of cyber adversity. With your updated IRP in hand, you'll be ready to channel your inner Will Smith and Jeff Goldblum, taking on whatever cyber challenges come your way. Now, let's get out there and save the digital world, one cyber incident at a time!

Trial Runs and Tune-ups
Prepping Your IRP for the Cyber Showdown of the Century

Much like how the pilots in 'Independence Day' had to practice their skills and adapt their strategies to defeat the alien invaders, it's crucial to test and update your IRP regularly. After all, you don't want to find out your plan has more holes than the Swiss cheese served at the victory party, do you? The following are how to ensure your IRP is battle-ready and prepared for any cyber showdown.

Tabletop Exercises: Gather your Cyber Avengers and conduct tabletop exercises to simulate various cyber incident scenarios. These simulations will help your team practice their response, identify gaps in your plan, and uncover opportunities for improvement. Remember, practice makes perfect—even Captain Steven Hiller (Will Smith) had to log some flight hours before saving the world!

Real-World Incident Analysis: Learn from the experiences of others—both their triumphs and their missteps. Analyze real-world cyber incidents to better understand how your organization might respond in similar situations. After all, hindsight is 20/20, and you don't want to end up like that one guy who thought he could take on an alien spacecraft with a crop duster.

Regular Reviews and Updates: The cyber threat landscape is constantly evolving, and so should your IRP. Schedule regular reviews to assess your plan's effectiveness, incorporating feedback from your team and any changes in your organization's environment. It's like keeping up with the latest alien-fighting tech or mastering new strategies in a high-stakes game of intergalactic chess.

External Audits: Sometimes, it takes an outsider's perspective to spot vulnerabilities you might have missed. Consider bringing in external experts to evaluate your IRP and offer recommendations for improvement. These experts can provide valuable insights and a fresh set of eyes, ensuring your plan is ready for the big leagues (or, you know, the next massive cyberattack).

By testing and updating your IRP regularly, you'll be well-prepared to handle any cyber crisis that comes your way. So, channel your inner Bill Pullman and lead your team to victory, knowing that your plan is as battle-tested and adaptable as the heroes of 'Independence Day.'

Drawing Wisdom from the Stars
When Fictional Heroes Teach Us Real-Life Lessons

Sometimes, the best teachers come in the most unexpected forms—like the brave heroes of 'Independence Day.' They might be fictional characters fighting otherworldly foes, but their experiences can still offer valuable insights and guidance for crafting an effective IRP. Let's take a closer look at some of these lessons and how they translate to the world of cybersecurity:

Embrace Unconventional Solutions: When faced with an overwhelming alien threat, David Levinson (played by Jeff Goldblum) came up with an ingenious plan to use a computer virus to bring down the invaders' shields. In cybersecurity, sometimes you need to think outside the box to counter sophisticated attacks. Don't be afraid to explore creative strategies and solutions—just make sure they're grounded in sound security practices.

Teamwork Makes the Dream Work: The human race triumphed in 'Independence Day' because everyone, from military personnel to computer experts, worked together toward a common goal. Similarly, a successful IRP requires collaboration from various departments within your organization. Foster a culture of teamwork and open communication, and you'll be better equipped to handle any cyber crisis.

Adapt and Overcome: In the face of adversity, the characters in 'Independence Day' learned to adapt their tactics and overcome seemingly insurmountable obstacles. Your IRP should be flexible and agile, able to adapt to the ever-changing cyber threat landscape. Keep an eye on emerging trends and new technologies, and don't hesitate to revise your plan as needed.

Never Give Up: Even when things seemed utterly hopeless, the heroes of 'Independence Day' refused to surrender. In cybersecurity, it's crucial to maintain a resilient mindset and be prepared to learn from setbacks. Remember, an incident response plan isn't a guarantee that you'll never experience a cyber incident—but it is an essential tool for minimizing the impact and bouncing back stronger than ever.

So, there you have it! By drawing inspiration from the (fictional) experts, you can craft an IRP that's as formidable and effective as the heroes of 'Independence Day.' Now, go forth and save the (digital) world!

Independence Day
Crafting Your Cyber Incident Response Plan

In this chapter, we focused on the vital process of creating an effective Incident Response Plan (IRP) using a casual, engaging writing style and drawing satirical inspiration from the movie 'Independence Day.' The chapter emphasized the importance of an IRP in guiding organizations through cyber incidents and minimizing their impact.

Throughout the chapter, we explored various aspects of IRP development, including the core components, preparing for the unexpected, and a step-by-step guide to crafting a comprehensive plan. We also discussed the importance of testing and updating the IRP and learning from both real-world and fictional experts.

By incorporating humor, real-world examples, and practical advice, we empowered readers with the knowledge and tools needed to create a robust IRP, equipping them to defend their digital assets against potential cyber threats effectively.

Cyber Security: The Final Frontier
A Journey of Digital Discovery, Laughter, and Triumph

As we reach the end of our cyber adventure, it's time to reflect on the wild ride we've been on together. From drawing inspiration from the worlds of StarCraft and 'Independence Day' to learning valuable lessons from Aunt Karen's potato salad and the antics of 'The Office,' we've explored the vast and ever-evolving landscape of cyber security with humor, wit, and a hearty dose of pop culture.

This journey has taught us that cyber security is more than just a collection of technical jargon and complex concepts—it's an essential part of our daily lives, a realm where creative thinking and collaboration can make all the difference. And, most importantly, we've learned that even in the face of daunting challenges, a little laughter can go a long way in making the digital world a safer and more enjoyable place.

As you embark on your own cyber security quests, armed with the knowledge and wisdom gleaned from these pages, remember that the journey never truly ends. The digital frontier is vast and ever-changing, filled with new challenges and opportunities waiting to be discovered. Stay curious, stay vigilant, and never stop learning.

So, dear reader, as we part ways, we leave you with one final thought: In the immortal words of President Thomas J. Whitmore from 'Independence Day,' "We will not go quietly into the night! We will not vanish without a fight! We're going to live on! We're going to survive!" And with that fighting spirit, let us continue our quest to create a safer, more secure digital world for all.

Godspeed, fellow cyber warriors!

Bibliography

- **Anderson, R. (2008). Security Engineering**: A Guide to Building Dependable Distributed Systems. Wiley Publishing.

- **Bejtlich, R. (2013). The Practice of Network Security Monitoring**: Understanding Incident Detection and Response. No Starch Press.

- **Chuvakin, A., & Schmidt, G. (2018). Security Operations Center**: Building, Operating, and Maintaining Your SOC. Cisco Press.

- Fruhlinger, J. (2018). The 18 biggest data breaches of the 21st century. CSO Online. Retrieved from https://www.csoonline.com/article/2130877/the-biggest-data-breaches-of-the-21st-century.html

- Grimes, R. (2019). 9 types of malware and how to recognize them. CSO Online. Retrieved from https://www.csoonline.com/article/3239673/9-types-of-malware-and-how-to-recognize-them.html

- Howard, J., & Longstaff, T. (1998). A Common Language for Computer Security Incidents. Sandia Report.

- Hutchins, E., Cloppert, M., & Amin, R. (2011). Intelligence-driven computer network defense informed by analysis of adversary campaigns and intrusion kill chains. Lockheed Martin Corporation.

- MITRE. (2021). ATT&CK. Retrieved from https://attack.mitre.org/

- National Institute of Standards and Technology. (2018). Framework for Improving Critical Infrastructure Cybersecurity. Retrieved from https://nvlpubs.nist.gov/nistpubs/CSWP/NIST.CSWP. 04162018.pdf

- Ritchey, R., & Amick, B. (2016). Network Security Bible. Wiley Publishing.

- Stuttard, D., & Pinto, M. (2011). The Web Application Hacker's Handbook: Finding and Exploiting Security Flaws. Wiley Publishing.

- Swauger, J. (2014). How to Write an Information Security Policy. SANS Institute. Retrieved from https://www.sans.org/reading-room/whitepapers/policyissues/write-information-security-policy-1791

- Tittel, E., & Chapple, M. (2015). CISSP For Dummies. Wiley Publishing.

- Verizon. (2021). Verizon Data Breach Investigations Report. Retrieved from https://enterprise.verizon.com/resources/reports/dbir/

- **StarCraft**: https://starcraft.com/en-gb/

- **Star Trek TNG**: https://ca.startrek.com/shows/star-trek-the-next-generation

- **Independence Day 1998**:
  https://www.imdb.com/title/tt0116629/

- **The Office USA**:
  https://www.imdb.com/title/tt0386676/

Acknowledgments

First and foremost, I want to express my deepest gratitude to Tom Moore, CEO of With You With Me, for his guidance, support, and the countless opportunities he has provided. Tom's vision and leadership have been instrumental in shaping my career, and I am truly grateful for the trust he has placed in me.

I would also like to thank Andy Choquette, a Solutions Architect I have had the privilege of working with on various projects. Andy's expertise, enthusiasm, and dedication have not only made our collaborations a success but also have been a source of inspiration for me. Your camaraderie and mentorship have made a lasting impact on my professional journey.

Last, but certainly not least, my heartfelt appreciation goes to my fiancée, Melissa Lariviere. Your unwavering love, encouragement, and patience have been the foundation upon which I have built my dreams. Melissa, your presence in my life has been a constant reminder of what truly matters, and this book would not have been possible without your steadfast support.

To everyone who has been a part of this journey, thank you for being a part of my story and for helping me bring this book to life.

About the Author

Travis Lothar Czech is a military veteran with 14 years of distinguished service under his belt. His passion for technology has led him to forge a successful career in Software Development, IT, and Cyber Security since the age of 18. As a seasoned professional, Travis currently holds the position of Sr. Tech Lead for a software engineering team, where he continues to make a significant impact in the industry.

In addition to his accomplished career, Travis is also a dedicated Cyber Security professor. With a focus on both offensive and defensive strategies, he is committed to educating the next generation of cyber security professionals, equipping them with the skills and knowledge needed to navigate the ever-evolving landscape of digital threats.

Drawing on his wealth of experience and expertise, Travis has authored this book to provide invaluable insights and guidance to those seeking to better understand and manage cyber security in their personal and professional lives. Through his unique perspective and engaging writing style, he aims to empower readers with the tools and information necessary to stay one step ahead in the world of cyber security.

www.ingramcontent.com/pod-product-compliance
Lightning Source LLC
Chambersburg PA
CBHW071026220526
45467CB00004B/1534